Mathematics pi edition

Write and Sketch 6" x 9" Journal

Writing Journal

Published by:
Berhampore Press
Wellington, NZ
Copyright 2017
All Rights Reserved

BerhamporePress@gmail.com

ISBN-13:
978-1545514894

ISBN-10:
1545514895

www.ingramcontent.com/pod-product-compliance
Lightning Source LLC
Chambersburg PA
CBHW071423180526
45170CB00001B/204